Headquarters
Department of the Army
Washington, DC
10 June 1987

Army Regulation 600–9

Effective 10 June 1987

Personnel—General

The Army Weight Control Program for all Personnel

By Order of the Secretary of the Army:

JOHN A. WICKHAM, JR.
General, United States Army
Chief of Staff

Official:

R. L. DILWORTH
Brigadier General, United States Army
The Adjutant General

History. The original form of this regulation was published on 1 September 1986. Since that time, changes have been issued to amend the original. As of 3 August 1987, permanent changes 1 and 2 remained in effect. This UPDATE printing incorporates all of those changes directly into the body of text. This publication has been reorganized to make it compatible with the Army electronic publishing database. No content has been changed.

Summary. This regulation implements guidance in DOD directive 1308.1, dated 29 June 1981, which establishes a weight control program in all the Services.

Applicability. This regulation applies to all members of the Active Army, the Army National Guard (ARNG), and the U.S. Army Reserve (USAR) to include those ARNG and USAR personnel in Active Guard/Reserve (AGR) status.

Proponent and exception authority. The proponent agency of this regulation is the Office of the Deputy Chief of Staff for Personnel.

Army management control process. This regulation is subject to the requirements of AR 11–2. It contains internal control provisions but does not contain checklists for conducting internal control reviews. These checklists will be contained in a DA Circular in the 11 series.

Supplementation. Supplementation of this regulation is prohibited without prior approval from HQDA (DAPE–MPA), WASH DC 20310.

Interim changes. Interim changes to this regulation are not official unless they are authenticated by The Adjutant General. Users will destroy interim changes on their expiration dates unless sooner superseded or rescinded.

Suggested Improvements. Users are invited to send comments and suggested improvements on DA Form 2028 (Recommended Changes to Publications and Blank Forms) directly to HQDA (DAPE–MPA), WASH DC 20310.

Distribution. Active Army, ARNG, and USAR: A.

Contents (Listed by paragraph and page number)

Contents—Continued

Appendixes

Table List

Figure List

Contents—Continued

Glossary

Section I
Introduction

1. Purpose
This regulation establishes policies and procedures for the implementation of the Army Weight Control Program.

2. References
Required publications are listed in appendix A.

3. Explanation of abbreviations and terms
Abbreviations and special terms used in this regulation are explained in the glossary.

4. Objectives
a. The primary objective of the Army Weight Control Program is to insure that all personnel—

(1) Are able to meet the physical demands of their duties under combat conditions.

(2) Present a trim military appearance at all times.

b. Excessive body fat—

(1) Connotes a lack of personal discipline.

(2) Detracts from military appearance.

(3) May indicate a poor state of health, physical fitness, or stamina.

c. Objectives of the Army Weight Control Program are to—

(1) Assist in establishing and maintaining—

(a) Discipline.

(b) Operational readiness.

(c) Optimal physical fitness.

(d) Health.

(e) Effectiveness of Army personnel through proper weight control.

(2) Establish appropriate body fat standards.

(3) Provide procedures for which personnel are counseled to assist in meeting the standards prescribed in this regulation.

(4) Foster high standards of professional military appearance expected of all personnel.

Section II
Responsibilities

5. General
The Army traditionally has fostered a military appearance which is neat and trim. Further, an essential function of day-to-day effectiveness and combat readiness of the Army is that all personnel are healthy and physically fit. Self-discipline to maintain proper weight distribution and high standards of appearance are essential to every individual in the Army.

6. Deputy Chief of Staff for Personnel (DCSPER)
The DCSPER is responsible for the Army Weight Control Program.

7. The Surgeon General (TSG)
TSG will—

a. Establish medical examination and counseling policies in support of the Army Weight Control Program.

b. Evaluate the medical aspects of the program.

c. Establish and review procedures for determination of body fat content.

d. Provide guidance on improving the nutritional status of soldiers.

8. Deputy Chief of Staff for Logistics (DCSLOG)
The DCSLOG will—

a. Establish food service guidance in support of the Army Weight Control Program.

b. Publish guidance and information pertaining to the caloric content of items served on master menus.

9. Chief, National Guard Bureau (CNGB)
The CNGB will—

a. Implement and monitor the Army Weight Control Program in the ARNG (including units).

b. Take appropriate action under guidance prescribed in this regulation.

10. Chief, Army Reserve (CAR)

The CAR will—

a. Monitor the Army Weight Control Program in the USAR.

b. Take appropriate action under guidance prescribed in this regulation.

11. Commanders of major Army commands (MACOMs)

MACOM commanders will insure that soldiers within their commands are evaluated under the body fat standards prescribed in this regulation.

12. Commanding General, US Army Training and Doctrine Command (CG, TRADOC)

The CG, TRADOC will inform personnel at initial entry on active duty (AD) about basic nutrition and sound food consumption practices.

13. Commanding General, US Army Forces Command (CG, FORSCOM)

The CG, FORSCOM will implement and maintain the Army Weight Control Program in USAR troop program units (TPUs).

14. Commanders of major medical commands

a. Commanding General, US Army Health Services Command (CG, HSC). The CG, HSC will—

(1) Institute weight reduction and counseling programs in Army medical facilities in support of the Army Weight Control Program.

(2) Provide appropriate literature and training aids for use by soldiers, supervisors, and commanders in selection of a proper diet.

b. Commanders of other major medical commands (overseas) will institute weight reduction and counseling programs in Army medical facilities in support of the Army Weight Control Program overseas.

15. Commanding General, US Army Reserve Components Personnel Center (CG, ARPERCEN)

The CG, ARPERCEN will—

a. Monitor the Army Weight Control Program in the Individual Ready Reserve (IRR).

b. Take appropriate action under guidance prescribed in this regulation.

c. Insure that members applying for tours of AD, active duty for training (ADT), active duty support (ADS), and AGR meet the body fat standards prescribed in this regulation. (Soldiers who do not meet these standards will not be permitted to enter on AD, ADT, or ADS, or in AGR status.)

16. Individuals

Each soldier (Commissioned Officer, Warrant Officer and Enlisted) is responsible for meeting the standards presecribed in this regulation. To assist soldiers in meeting these responsibilities, screening tables will continue to be prescribed for use as is currently being done. A five percent zone below the screening table weight ceiling is suggested as a help to soldiers in targeting their personal weight at a level which will minimize the probability of exceeding the screening table weight ceiling as a matter of habit. Soldiers should be coached to select their personal weight goal within or below the five percent zone and strive to maintain that weight through adjustment of life style and fitness routines. If a soldier consistently exceeds the personal weight goal, he or she should seek the assistance of master fitness trainers for advice in proper exercise and fitness; and health care personnel for a proper dietary program. In other words, exceeding a properly selected goal should "trigger" the soldier to use the substantial help available to alter the fitness and dietary behavior before confronting the finality of the screening table and initiation of official action if the body fat standards are exceeded.

17. Commanders and superviors

Commanders and supervisors (Active and Reserve Components) will—

a. Implement the Army Weight Control Program, to include evaluation of the weight and military appearance of all soldiers under their jurisdiction; to include measuring body fat as prescribed in this regulation.

b. Insure the continued evaluation of all soldiers under their command or supervision against the body fat standards prescribed in this regulation.

c. Maintain data on soldiers in their command or under their supervision who—

(1) Enter a weight control program each year.

(2) Subsequently either meet the body fat standards prescribed in this regulation or were separated from the Service for reasons related to overweight conditions.

d. Encourage soldiers to establish a personal weight goal as described above.

e. ensure that personnel responsible for issuing TDY and PCS orders include in the text of all orders the words: "You are responsible for reporting to your next duty station/school in satisfactory physical condition, able to pass the APFT and meet weight standards."

18. Health Care Personnel

Health care personnel will—

a. Assist commanders and supervisors by providing weight reduction counseling to individuals who are overweight.

b. Identify those individuals who have a pathological condition requiring medical treatment.

c. Evaluate overweight soldiers, when—

(1) A soldier has a medical limitation (DA Form 3349, AR 40–501 prescribes assignment limitations for soldiers with profiles. For example, no mandatory strenuous physical activity); or

(2) Is pregnant; or

(3) When an evaluation is requested by a unit commander (this is an option for unit commanders and is not mandatory); or

(4) Separation is being considered for failure to make satisfactory progress in a weight control program; or

(5) Six months prior to ETS.

19. Master Fitness Trainers

Master Fitness Trainers will—

a. Prescribe proper exercise and fitness techniques to assist soldiers in determining, achieving and maintaining an appropriate personal weight goal.

b. Assist commanders in developing proactive programs which clearly establish physical fitness as a unit value.

Section III
Proper Weight Control

20. Policy

a. Commanders and supervisors will monitor all members of their command (officers, warrant officers, and enlisted personnel) to insure that they maintain proper weight, body composition (as explained in the glossary), and personal appearance. At minimum, personnel will be weighed when they take the Army Physical Fitness Test (APFT) or at least every 6 months. Personnel exceeding the screening table weight (shown in table 1) or identified by the commander or supervisor for a special evaluation will have a determination made of percent body fat. Identification and counseling of overweight personnel are required.

b. Commanders and supervisors will provide educational and other motivational programs to encourage personnel to attain and maintain proper weight (body fat) standards. Such programs will include—

(1) Nutrition education sessions conducted by qualified health care personnel. These sessions are required for all soldiers enrolled in a Weight Control Program.

(2) Exercise programs, even though minimum APFT standards are achieved.

c. Maximum allowable percent body fat standards are as follows in table 2. However, all personnel are encouraged to achieve the more stringent Department of Defense (DOD)-wide goal, which is 20 percent body fat for males and 26 percent body fat for females.

Table 2
Maximum allowable percent body fat standards

Age Group	Male (% body fat)	Female (% body fat)
17–20	%20	%28
21–27	%22	%30
28–39	%24	%32
40 & Older	%26	%34

d. Personnel who are overweight (as explained in the glossary) (including soldiers who become pregnant while on the weight control program)—

(1) Will be considered nonpromotable (to the extent such nonpromotion is permitted by law).

(2) Will not be authorized to attend professional military or civilian schooling. Personnel arriving at professional military schools overweight will be denied enrollment, except:

(a) for personnel in a TDY and return or a TDY enroute to PCS status who, in the opinion of the appropriate school commander or commandant, can potentially meet their body fat standard by the end of the course or within 30 days, whichever comes first. Overweight personnel who do not meet their body fat standard within the probationary period will be disenrolled and will not graduate. An academic efficiency report as required, will be annotated to indicate failure to meet body fat standards, and the appropriate career division will be notified for those personnel in a TDY enroute to PCS status.

(b) for personnel in a PCS status, and then, only for a 30 day probationary period. If at the end of the probationary period, body fat standards have not been met, the student will be disenrolled. An academic efficiency report as required, will be annotated to indicate failure to meet body fat standards, and the appropriate career division will be notified for assignment instructions.

(3) Will not be assigned to command positions.

21. Procedures

a. Body fat composition will be determined for personnel—

(1) Whose body weight exceeds the screening table weight in table 1.

(2) When the unit commander or supervisor determines that the individual's appearance suggests that body fat is excessive.

b. Routine weigh-ins will be accomplished at the unit level. Percent body fat measurements will be accomplished by company or similar level commanders (or their designee) in accordance with standard methods prescribed in Appendix B to this regulation. Soldiers will be measured by individuals of the same gender. If this cannot be accomplished, a female soldier will be present when males measure females. IRR members on AT, ADT, and SADT will have a weigh-in, and body fat evaluation (if required) by the unit to which attached. Active and Reserve component soldiers exceeding the body fat standards in paragraph 20*c*, above, will be provided weight reduction counseling by health care personnel, entered in a Weight Control Program by unit commanders, and flagged under the provisions of AR 600–31 by the unit commander. Enrollment in a Weight Control Program starts on the day that the soldier is informed by the unit commander that he/she has been entered in a weight control program. The weight reduction counseling can be accomplished prior to or shortly after entry into a program.

c. The sample correspondence shown in figure 1 will be completed and retained by the unit commander or supervisor to document properly recommendations and actions taken in each case.

d. A medical evaluation will be accomplished by health care personnel when the soldier has a medical limitation, or is pregnant, or when requested by the unit commander. One is also required for soldiers being considered for separation due to failure to make satisfactory progress in a weight control program, or within 6 months of ETS. Aircraft crew members, who exceed the body fat standards, will be referred to a flight surgeon for possible impact on flight status. If an individual's condition is diagnosed by medical authorities to result from an underlying or associated disease process, health care personnel will take one of the following actions:

(1) Prescribe treatment to alleviate the condition and return personnel to their unit.

(2) Hospitalize individuals for necessary treatment; this action applies to Active Army personnel only. Reserve Component personnel will be referred to their personal physician for further evaluation or treatment at the individual's expense.

(3) Determine whether the individual's condition is medically disqualifying for continued service. In these cases, disposition will be made under provisions of appropriate regulations.

e. If health care personnel discover no underlying or associated disease process as the cause of the condition (or RC personnel decline to be evaluated under paragraph 21*d*) and the individual is classified as overweight, these facts will be documented and the individual entered in a weight control program by the unit commander. Suspension of favorable personnel actions will be initiated under AR 600–31 for personnel in a weight control program.

(1) The required weight loss goal of 3 to 8 pounds per month is considered a safely attainable goal to enable soldiers to lose excess body fat and meet the body fat standards described in para 20*c*. Weight-ins will be made by unit personnel monthly (or during unit assemblies for ARNG and USAR personnel) to measure progress. A body fat evaluation may also be done by unit personnel to assist in measuring progress.

(2) As an exception to *g* below, an individual who has not made satisfactory progress after any two consecutive monthly weigh-ins may be referred by the commander or supervisor to health care personnel for evaluation or reevaluation. If health care personnel are unable to determine a medical reason for lack of weight loss—and if the individual is not in compliance with the body fat standards at paragraph 20*c* and still exceeds the screening table weight (table 1)—the commander or supervisor will inform the individual that—

(a) Progress is unsatisfactory.

(*b*) He or she is subject to separation as specified in *j* below.

f. Commanders and supervisors will remove individuals administratively from a weight control program as soon as the body fat standard is achieved. The screening table weight will not be used to remove soldiers from a Weight Control Program. The removal action will be documented as shown in figure 1; removal of suspension of favorable personnel actions will be accomplished at that time.

g. After a period of dieting and/or exercise for 6 months, soldiers who have not made satisfactory progress (as explained in the glossary) and who still exceed the screening table and body fat standards will be processed as follows:

(1) If health care personnel determine that the condition is due to an underlying or associated disease process, action described in *d* above will be taken.

(2) If no underlying or associated disease process is found to cause the overweight condition, the individual will be subject to separation from the Service under appropriate regulations indicated in *j* below.

h. Personnel will be continued in a weight control program (as provided in *e* through *g* above) after the initial 6-month period if they—

(1) Still exceed the body fat standard.

(2) Have made satisfactory progress toward their weight loss (as indicated in the glossary), or are at or below the screening table weight (table 1).

(3) *For RC personnel only.* If the individual has not obtained an evaluation from his/her personal physician under the provisions of 21*d* above, and cannot demonstrate that the overweight condition results from an underlying or associated disease process, the individual may be separated under appropriate regulations without further medical evaluation by health care personnel.

i. To assist commanders and supervisors, a flow chart outlining procedural guidance is shown at figure 2.

j. The commander or supervisor will inform the individual in writing that initiation of separation proceedings is being considered under the following regulations: AR 635–200, chapter 5–15, AR 635–100, chapter 5; NGR 600–200, chapter 7; NGR 600–101; NGR 600–5; NGR 635–100; AR 135–175; or AR 135–178. This procedure will be followed unless a medical reason is found to preclude the loss of weight or there is other good cause to justify additional time in the weight control program.

(1) The individual will immediately respond to the separation consideration letter in writing. The commander or supervisor will consider the response and initiate separation action if no adequate explanation is provided, unless the individual submits an application for retirement, if eligible. USAR personnelwho fall under the purview of this paragraph will be released from AD or discharged, as appropriate.

(2) If separation action is not initiated or does not result in separation, the individual will be entered or continued in a weight control program, as specified in *e* above.

k. Following removal from a weight control program, if it is determined (under *a* above) that an individual exceeds the screening table weight (table 1) and the body fat standard prescribed in paragraph 20*c* within 36 months, the following will apply:

(1) If the unit commander determines that the individual exceeds the screening table weight and the body fat standard—

(*a*) Within 12 months from the date of the previous removal from the program and no underlying or associated disease process is found as the cause of the condition, the individual will be subject to separation from the Service under *j* above.

(*b*) After the 12th month, but within 36 months from the date of the previous removal from the program, and no underlying or associated disease process is found as the cause of the condition, the individual will be allowed 90 days to meet the standards. Personnel who meet the body fat standard after that period will be removed from the program. All others will be subject to separation from the Service under *j* above.

(*c*) Personnel who meet the AR 600–9 standards and become pregnant will be exempt from the standards for the duration of the pregnancy plus the period of convalescent leave after termination of pregnancy. They will be entered in a weight control program, if required, after completion of convalescent leave and approval of a medical doctor that they are fit for participation in a weight control program. This procedure also applies to individuals in a medical holding unit who have been hospitalized for long periods. Soldiers entered/reentered in a weight control program after pregnancy, prolonged treatment, or hospitalization will be considered to be in a new weight control program. Paragraph 21*k* of this regulation will not apply at that time.

(2) If the individual is determined to exceed the body fat standard and the condition is due to an underlying or associated disease process, action described in *d* above will be taken.

l. Inherent in the responsibility of selection boards is the obligation to select only those individuals who are considered to be physically fit to perform the duties required of them at all times. Compliance with the Army Weight Control Program as prescribed in this regulation will be considered in the selection process for promotion, professional military or civilian schooling, or assignment to command positions. Procedures for commanders and supervisors to provide current information for use by selection boards indicating whether individuals meet the prescribed standards will be included in DA regulations or issued by separate correspondence.

m. Records will be maintained in the MPRJ for personnel in weight control programs. On transfer from one unit to another, the losing commander or supervisor will forward a statement to the gaining unit with information indicating the status of the individual's participation in a weight control program. When the transfer is a permanent change of station, the unit commander's statement will be filed as a transfer documents in the Military Personnel Records Jacket, U.S. Army, under AR 640–10.

n. Upon removal from the weight control program, unit records on participation in a weight control program will be maintained in the MPRJ for a period of 36 months from date of removal. If the soldier is transferred to another unit prior to completion of 36 months, action will be taken in accordance with paragraph 21*m* above.

22. Reenlistment criteria

a. Personnel who exceed the screening table weight at table 1 and the body fat standard for their current age group in paragraph 20*c* will not be allowed to reenlist or extend their enlistment.

b. Exceptions to policy for Active Army personnel (including RC personnel on AD) are prescribed in this subparagraph. For soldiers who are otherwise physically fit and have performed their duties in a satisfactory manner, the commander exercising General Court Martial Convening Authority or the first general officer in the soldier's normal chain of command (whichever is in the most direct line to the soldier) may approve the following exceptions to policy:

(1) Extension of enlistment may be authorized for personnel who meet one of the following criteria:

(a) Individuals who have a temporary medical condition which precludes loss of weight. In such cases, the nature of on-going treatment will be documented; the extension will be for the minimum time necessary to correct the condition and achieve the required weight loss.

(b) Pregnant soldiers (except those soldiers who have medical conditions as listed in para 21*d*) who are otherwise fully qualified for reenlistment, including those with approved waivers, but who exceed acceptable standards prescribed in this regulation, will be extended for the minimum period which will allow birth of the child, plus 6 months. A clearance from the doctor tht the soldier is medically fit to participate in a Weight Control Program is required. Authority, which will be cited on DA Form 1695 (Oath of Extension of Enlistment), is AR 601–280, paragraph 3–3. On completion of the period of extension, the soldier will be reevaluated under paragraph 21.

(2) Exceptions to policy allowing reenlistment/extension of enlistment are authorized only in cases where—

(a) Medically documented conditions (para 21*d*) preclude attainment of required standards and

(b) Disability separation is not appropriate.

c. All requests for extension of enlistment for ARNG and USAR (TPU and IRR) personnel not on AD will be processed under NGR 600–200 or AR 140–111, chapter 3, as appropriate.

d. Requests for exceptions to policy will be forwarded through the chain of command, with the commander's personal recommendation and appropriate comment at each level. As a minimum, requests will include—

(1) The physician's evaluation.

(2) A record of progress in the weight control program.

(3) Current height and weight.

(4) Body fat content.

(5) Years of active Federal service.

(6) Other pertinent information.

e. Soldiers, who have completed a minimum of 18 years of Active Federal Service (AFS), may, if otherwise eligible, be extended for the minimum time required to complete 20 years AFS. Retirement must be accomplished not later than the last day of the month in which the soldier attains retirement eligibility. Application for retirement will be submitted at the time extension is authorized. Approval/disapproval authority is outlined in AR 601–280.

f. USAR soldiers, who have completed a minimum of 18 years of qualifying service for retired pay at age 60, may be extended for the minimum time required to complete 20 years qualifying service. Approval/disapproval authority is outlined in AR 140–111. Transfer to the IRR, Retired Reserve, or discharge will be accomplished at the end of the retirement year (RYE) in which the soldier attains the 20 qualifying years.

g. ARNG soldiers, who have completed a minimum of 18 years qualifying service for retired pay at age 60, may be extended for the minimum time required to complete 20 years qualifying service by the State TAG, disapproval authority is C, NGB. Transfer to the IRR, Retired Reserve, or discharge will be accomplished at the end of the retired year (RYE) in which the soldier attains the 20 qualifying years.

Table 1
Weight for Height Table (Screening Table Weight)

Height (in inches)	Male Age				Female Age			
	17–20	21–27	28–39	40+	17–20	21–27	28–39	40+
58	—	—	—	—	109	112	115	119
59	—	—	—	—	113	116	119	123
60	132	136	139	141	116	120	123	127
61	136	140	144	146	120	124	127	131
62	141	144	148	150	125	129	132	137
63	145	149	153	155	129	133	137	141
64	150	154	158	160	133	137	141	145
65	155	159	163	165	137	141	145	149
66	160	163	168	170	141	146	150	154
67	165	169	174	176	145	149	154	159
68	170	174	179	181	150	154	159	164
69	175	179	184	186	154	158	163	168
70	180	185	189	192	159	163	168	173
71	185	189	194	197	163	167	172	177
72	190	195	200	203	167	172	177	183
73	195	200	205	208	172	177	182	188
74	201	206	211	214	178	183	189	194
75	206	212	217	220	183	188	194	200
76	212	217	223	226	189	194	200	206
77	218	223	229	232	193	199	205	211
78	223	229	235	238	198	204	210	216
79	229	235	241	244	203	209	215	222
80	234	240	247	250	208	214	220	227

Notes:

1. The height will be measured in stocking feet (without shoes), standing on a flat surface with the chin parallel to the floor. The body should be straight but not rigid, similar to the position of attention. The measurement will be rounded to the nearest inch with the following guidelines:
   a. If the height fraction is less than 1/2 inch, round down to the nearest whole number in inches.
   b. If the height fraction is 1/2 inch or greater, round up to the next highest whole number in inches.
2. The weight should be measured and recorded to the nearest pound within the following guidelines.
   a. If the weight fraction is less than 1/2 pound, round down to the nearest pound.
   b. If the weight fraction is 1/2 pound or greater, round up to the next highest pound.
3. All measurements will be in a standard PT uniform (gym shorts and T-shirt, without shoes).
4. If the circumstances preclude weighing soldiers during the APFT, they should be weighed within 30 days of the APFT.
5. Add 6 pounds per inch for males over 80 inches and 5 pounds for females for each inch over 80 inches.

(Letterhead)

Office Symbol **Date**

SUBJECT: Weight Control Program

(Individual Soldier)

1. You have been determined to exceed the body fat standard and a goal of 3-8 pounds of weight loss per month is considered to be satisfactory progress. Failure to make satisfactory progress or achieve the body fat standards could result in separation from the service.

2. You have been flagged under the provisions of AR 600–31 and entered in a Weight Control Program.

Signature of Unit Commander

Office Symbol 1st End **Date**

FROM: Individual Soldier

TO: Commanding Officer

I understand my responsibilities to achieve the body fat standards to have my weight recorded periodically or during unit training assemblies (UTA) as applicable.

Soldier's Signature

Office Symbol 2d End

SUBJECT: Weight Control Program

FROM: Unit Commander

TO: MEDDAC

1. . . . exceeds the weight for height tables by . . . pounds and exceeds the body fat standards by .percent.

2. It is requested that a medical evaluation be conducted in view of the following (check applicable block):

( )Soldier's profile. ( )Pregnancy. ( )Unit Commander's special request. ( )Initiation of separation action (failure to make satisfactory progress in a Weight Control Program). ( )Within 6 months of ETS.

Signature of Unit Commander

Office Symbol 3d End

SUBJECT: Weight Control Program

FROM: Health Care Personnel

TO: Commanding Officer

1. In accordance with AR 600–9 . . . has been examined and found to be fit for participation in a Weight Control/Physical Exercise Program.

2. The cause of the overweight condition (is) (is not) due to a medical condition.

3. The following action(s) is/are recommended:

( )Initiation or continuation in a weight reduction program.

( )Medical treatment for pathological medical disorder (refer Reserve Component soldiers to their personal physician for evaluation at their own expense).

Signature of Health Care Personnel

Figure 1. Sample correspondence for Weight Control Program

Office Symbol 4th End

FROM: Unit Commander

TO: MEDDAC (Health Care Personnel)

1. . . . exceeds the screening table weight by . . . and body fat standards by . . . percent.

2. Nutrition education and weight reduction counseling are requested in accordance with AR 600–9 (para 20b (1)).

Signature of Unit Commander

Office Symbol 5th End

SUBJECT: Weight Control Program

FROM: Health Care Personnel

TO: Commanding Officer

1. . . . has been provided nutrition and weight reduction counseling in accordance with AR 600–9.

2. Follow-up counseling should be provided at unit level using information in Appendix C of this regulation and the assistance of master fitness trainers, if available.

Signature of Health Care Personnel

Office Symbol 6th End

FROM: Unit Commander

TO: Military Personnel Officer

1. The soldier has been determined to be in compliance with the provisions of AR 600–9, and is therefore removed from the Weight Control Program effective this date.

2. The individual's current weight is . . . pounds. Screening table weight ceiling is . . . pounds for present age category. Body fat content is . . . percent, which is within the AR 600–9 standards.

3. This correspondence will be retained in the individual's MPRJ for 36 months from this date.

Signature of Unit Commander

Figure 1. Sample correspondence for Weight Control Program—Continued

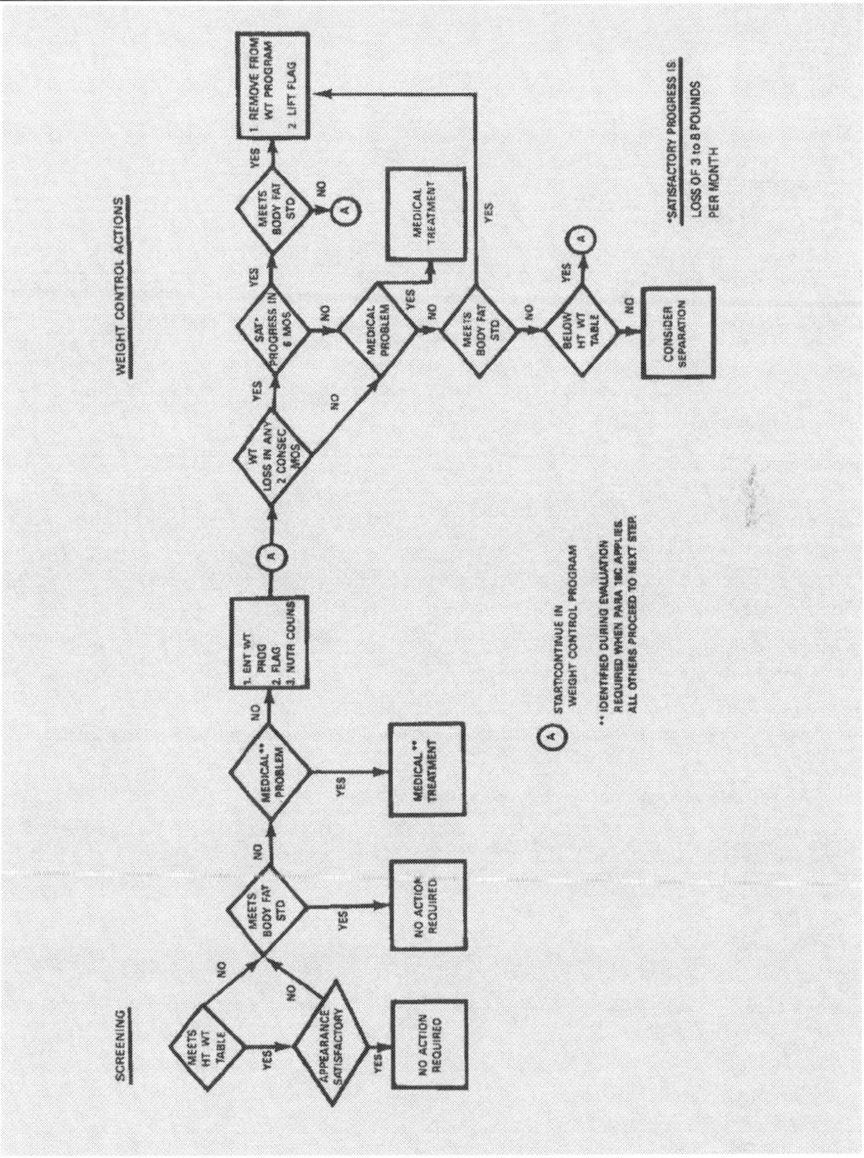

Figure 2. Flow process guide for screening and weight control actions (applies to all personnel; officers, warrant officers, enlisted)

Appendix A
References

Section I
Required Publications

AR 135–175
Separation of Officers. (Cited in paragraph 20*j*.)

AR 135–178
Separation of Enlisted Personnel. (Cited in paragraph 20*j*.)

AR 140–111
Enlistment and Reenlistment. (Cited in paragraph 21*c*.)

AR 600–31
Suspension of Favorable Personnel Actions for Military Personnel in National Security Cases and Other Investigations or Proceedings. (Cited in paragraph 20*e*.)

AR 601–280
Army Reenlistment Program. (Cited in paragraph 21.)

AR 635–100
Officer Personnel. (Cited in paragraph 20*j*.)

AR 635–200
Enlisted Personnel. (Cited in paragraph 20*j*.)

AR 640–10
Individual Military Personnel Records. (Cited in paragraph 20*m*(1).)

NGR 600–5
Enlisted Career Management. (Cited in paragraph 20*j*.)

NGR 600–101
Commissioned Officers—Federal Recognition and Related Personnel Activities. (Cited in paragraph 20*j*.)

NGR 600–200
Enlisted Personnel Management. (Cited in paragraphs 20*j* and 21*d*.)

NGR 635–100
Termination of Appointment and Withdrawal of Federal Recognition. (Cited in paragraph 20*j*.)

Section II
Related Publications
This section contains no entries.

Section III
Prescribed Forms
This section contains no entries.

Section IV
Referenced Forms
This section contains no entries.

Appendix B
Standard Methods for Determining Body Fat Using Body Circumferences, Height and Weight

B–1. Introduction

a. The procedures for the measurements of height, weight, and specific body circumferences for the estimation of body fat are described in this appendix.

b. Although circumferences may be looked upon by untrained personnel as easy measures, they can give erroneous results if proper precautions are not followed. The individual taking the measurements must have a thorough understanding of the appropriate body landmarks and measurement techniques. Unit commanders should require that designated personnel have hands-on training and read the instructions regarding technique and location, and practice before official determinations are made. Two members of the unit should be utilized in the taking of measurements, one to place the tape measure and determine measurements, the other to assure proper placement and tension of the tape, as well as to record the measurement on the worksheet. The individual taking the measurements should be of the same sex as the soldier being measured; the individual who assists the measurer and does the recording may be of either sex. The two should work with the soldier between them so the tape is clearly visible from all sides. Measurements will be made three times, in accordance with standard body measurement procedures. This is necessary for reliability purposes, since the greater number of measurements, the lesser the standard of deviation. Also, if only two measurements were taken, there would be no way to tell which measurement was the most accurate. If there is greater than 1/4-inch difference between the measurements, then continue measuring until you have three measurements within 1/4-inch of each other. An average of the scores that are within 1/4-inch of each other will be used.

c. When measuring circumferences, compression of the soft tissue is a problem that requires constant attention. The tape will be applied so that it makes contact with the skin and conforms to the body surface being measured. It should not compress the underlying soft tissues. Note, however, that in the hip circumference more firm pressure is needed to compress gym shorts. All measurements are made in the horizontal plane, (i.e., parallel to the floor), unless indicated otherwise.

d. The tape measure should be made of a non-stretchable material, preferably fiberglass; cloth or steel tapes are unacceptable. Cloth measuring tapes will stretch with usage and most steel tapes do not conform to body surfaces. The tape measure should be calibrated, i.e., compared with a yardstick or a metal ruler to ensure validity. This is done by aligning the fiberglass tape measure with the quarter inch markings on the ruler. The markings should match those on the ruler; if not, do not use that tape measure. The tape should be 1/4- to 1/2-inch wide (not exceeding 1/2-inch) and a minimum of 5–6 feet in length. A retractable fiberglass tape is the best type for measuring all areas. Tapes currently available through the Army Supply System (Federal Stock Number 8315–00–782–3520) may exceed the 1/2-inch width limits and could slightly impact on circumferential measurements. Efforts are being made to replace the supply system tape with a narrower retractable tape. In the interim, the current Army supply system or any other fiberglass tape not to exceed 5/8-inch may be used if retractable tapes cannot be purchased by unit budget funds available and approved by installation commanders.

B–2. Height and weight measurements

a. The height will be measured with the soldier, in stocking feet (without shoes) and standard PT uniform, i.e., gym shorts and T- shirt, standing on a flat surface with the head held horizontal, looking directly forward with the line of vision horizontal, and the chin parallel to the floor. The body should be straight but not rigid, similar to the position of attention. Unlike the screening table weight this measurement will be recorded to the nearest 1/4-inch in order to gather a more accurate description of the soldier's physical characteristics.

b. The weight will be measured with the soldier in a standard PT uniform, i.e., gym shorts and a T-shirt. Shoes will not be worn. The measurement should be made on scales available in units and recorded to the nearest pound with the following guidelines:

(1) If the weight fraction of the soldier is less than 1/2-pound, round down to the nearest pound.

(2) If the weight fraction of the soldier is 1/2-pound or greater, round up to the next whole pound.

B–3. Description of circumference sites, and their anatomical landmarks and technique

a. All circumference measurements will be taken three times and recorded to the nearest 1/4-inch (or 0.25). Each sequential measurement should be within 1/4-inch of the next or previous measurement. If the measurements are within 1/4-inch of each other, derive a mathematical average to the nearest quarter (1/4) of an inch. If the measurements differ by 1/4-inch or more continue measurements until you obtain three measures within 1/4-inch of each other. Then average the three closest measures.

b. Each set of measurements will be completed sequentially to discourage assumption of repeated measurement readings. For males, complete 1 set of abdomen and neck measurements, NOT three abdomen circumferences followed by three neck circumferences. Continue the process by measuring the abdomen and neck in series until you have three sets of measurements. For females, complete one set of hip, forearm, neck, and wrist measurements, NOT 3 hip followed by three forearm etc. continue the process by measuring hip, forearm, neck, and wrist series until you have 3 sets of measurements.

c. Worksheets for computing body fat are at figure B–1 (males) and figure B–3 (females). Local reproduction is authorized. A blank copy of DA Forms 5500–R and 5501–R is located at the back of this volume. These forms will be reproduced locally on 8½ x 11-inch paper. Supporting factor tables are located at tables B–1 and B–2 (males) and tables B–3 through B–8 (females) and include specific steps for preparing body fat content worksheets.

d. Illustrations of each tape measurement are at figure B–2 (males) and figure B–4 (females). A training videotape (TVT 8–103) is also available at Visual Information Libraries, and/or Training Audiovisual Support Centers (TASC).

B–4. Circumference sites and landmarks for males

a. Abdomen. The soldier being measured will be standing with arms relaxed. The abdominal measurement is taken at a level coinciding with the midpoint of the navel (belly button) with the tape placed so that it is level all the way around the soldier being measured. Record the measurement at the end of a normal expiration. It is important that the soldier does not attempt to hold his abdomen in, thus resulting in a smaller measurement. Also the tape must be kept level across the abdomen and back.

b. Neck. The soldier being measured will be standing, looking straight ahead, chin parallel to the floor. The measurement is taken by placing the tape around the neck at a level just below the larynx (Adam's apple). Do not place the tape measure over the Adam's apple. The tape will be as close to horizontal (the tape line in the front of the neck should be at the same height as the tape line in the back of the neck) as anatomically feasible. In many cases the tape will slant down toward the front of the neck. Therefore, care should be taken so as not to involve the shoulder/ neck muscles (trapezius) in the measurement. This is a possibility when a soldier has a short neck.

B–5. Circumference sites and landmarks for females

a. Neck. This procedure is the same as for males.

b. Forearm. The soldier being measured will be standing with the arm extended away from the body so that the forearm is in plain view of the measurer, with the hand palm up. The soldier should be allowed to choose which arm he/she prefers to be measured. Place the tape around the largest forearm circumference. This will be just below the elbow. To ensure that this is truly the largest circumference, since it is being visually identified, slide the tape along the forearm to find the largest circumference.

c. Wrist. The soldier being measured will stand with the arm extended away from the body so that the wrist is in plain view of the measurer. The tape will be placed around the wrist at a point above the hand just below the lower end of the bones of the forearm.

d. Hip. The soldier taking the measurement will view the person being measured from the side. Place the tape around the hips so that it passes over the greatest protrusion of the gluteal muscles (buttocks) keeping the tape in a horizontal plane (i.e., parallel to the floor). Check front to back and side to side to be sure the tape is level to the floor on all sides before the measurements are recorded. Since the soldier will be wearing gym shorts, the tape can be drawn snugly to minimize the influence of the shorts on the size of the measurement.

B–6. Preparation of the body fat content worksheets

NOTE: IT IS EXTREMELY IMPORTANT THAT YOU READ ALL OF THESE INSTRUCTIONS BEFORE ATTEMPTING TO COMPLETE THE BODY FAT CONTENT WORKSHEETS. MAKE SURE THAT YOU HAVE A COPY OF THE WORKSHEET IN FRONT OF YOU WHEN YOU ARE READING THESE INSTRUCTIONS.

a. The following paragraphs will provide information needed to perpare the Body Fat Content Worksheets for males and females, DA Form 5500–R and 5501–R, Dec 85. The worksheets are written in a stepwise fashion. The measurements and computation processes are different for males and females.

b. You will be responsible for completing a worksheet for soldiers who exceed the screening table weight (Table 1) located in this regulation, or when a unit commander or supervisor determines that the individual's appearance suggests that body fat is excessive (para 20*a* AR 600–9). The purpose of this form is to help you determine the soldier's percent body fat using the circumference technique described in this regulation.

c. Before you start, you should have a thorough understanding of the measurements to be made as outlined in this appendix. You will also need a scale for measuring body weight, a height measuring device, and a measuring tape (see specifications in para B–1*d*) for the circumference measurements.

B–7. Steps for preparing the Male Body Fat Content Worksheet, DA Form 5500–R, Dec 85

Name Print the soldier's last name, first name, and middle initial in the NAME block. Also include his Rank, and Social Security Number.

Age Print his age in years in the AGE block.

Height Measure the soldier's height as described in this appendix, to the nearest quarter of an inch, and record the measurement in the HEIGHT block.

Weight Measure the soldier's weight as described in this appendix, to the nearest pound, and record in the WEIGHT block.

Note: Follow the rules for rounding of height and weight measurements as described earlier in this appendix.

Step 1. Abdominal Measurement

Measure the soldier's abdominal circumference to the nearest quarter of an inch, and record in the block labeled "FIRST".

Step 2. Neck Measurement

Measure the soldier's neck circumference to the nearest quarter of an inch, and record in the block labeled "FIRST".

Note: REPEAT STEPS 1 and 2, in series until you have completed three sets of abdomen and neck circumferences.

Step 3. Average Abdominal Measurement

Find the mathematical average of the FIRST, SECOND, and THIRD abdominal circumferences by adding them together and dividing by three. Place this number to the nearest quarter of an inch, in the block marked AVERAGE, for STEPS 1 and 3.

Step 4. Average Neck Measurement

Find the mathematical average of your FIRST, SECOND, and THIRD neck circumferences by adding them together and dividing by three. Place this number to the nearest quarter of an inch in the block marked AVERAGE, for STEPS 2 and 4.

Step 5. Abdomen-Neck Difference

Subtract the number found in the AVERAGE block of STEP 4 from the number found in the AVERAGE block in STEP 3. Enter the result in STEP 5. This is the difference between the abdomen and neck circumferences.

Step 6. Abdomen-Neck Factor

Go to Table B–1, the Abdomen-Neck Factor Table, and locate the abdomen-neck difference in the left-most column. If the difference is a whole number, i.e., 15 inches, the Abdomen-Neck Factor is 89.93. If the difference is 15.25 inches, the factor would be 90.48, if the difference is 15.50, the factor is 91.02, and if the difference is 15.75, the factor is 91.55. Enter the appropriate factor in STEP 6.

Step 7. Height Factor

Go to Table B–2, the Height Factor Table, and locate the soldier's height in the left-most column. If the height is a whole number, i.e., 64 inches, the factor is 77.15. If the height is not a whole number, i.e., 64.25 inches, the factor is 77.27, if the height is 64.50 inches the factor is 77.39, and if the height is 64.75 inches, the factor is 77.50. Enter the appropriate factor in STEP 7.

Note: Therefore, the general rule for the factor tables is if the measure or difference is a whole number, your factor will be located under the 0.00 column, directly across from the inches column. Columns .25, .50, .75 correspond to measurements that are not whole numbers, but rather are fractions of an inch.

Step 8. Percent Body Fat

Subtract the number found in the AVERAGE block of STEP 7 from the number found in the AVERAGE block of STEP 6 and enter the difference in STEP 8. This is the soldier's PERCENT BODY FAT.

Table B–1
Male Abdomen and Neck Factor

Difference in inches	0.00	.25 (¼)	.50 (½)	.75 (¾)
5	53.44	55.06	56.61	58.09
6	59.50	60.85	62.16	63.41
7	64.62	65.78	66.91	68.00
8	69.05	70.07	71.07	72.03
9	72.96	73.87	74.76	75.62
10	76.46	77.28	78.08	78.86
11	79.63	80.37	81.10	81.82
12	82.52	83.20	83.87	84.53
13	85.17	85.81	86.43	87.04
14	87.64	88.22	88.80	89.37
15	89.93	90.48	91.02	91.55
16	92.07	92.58	93.09	93.59
17	94.08	94.57	95.05	95.52
18	95.98	96.44	96.89	97.34
19	97.78	98.21	98.64	99.06
20	99.48	99.89	100.30	100.70
21	101.10	101.49	101.88	102.26
22	102.64	103.02	103.39	103.76
23	104.12	104.48	104.83	105.19
24	105.53	105.88	106.22	106.56
25	106.89	107.22	107.55	107.87
26	108.19	108.51	108.82	109.14
27	109.44	109.75	110.05	110.35
28	110.65	110.95	111.24	111.53
29	111.82	112.10	112.39	112.67
30	112.94	113.22	113.49	113.76
31	114.03	114.30	114.56	114.83
32	115.09	115.35	115.60	115.86
33	116.11	116.36	116.61	116.85
34	117.10	117.34	117.58	117.82
35	118.06	118.30	118.53	118.77
36	119.00	119.23	119.46	119.68
37	119.91	120.13	120.35	120.57
38	120.79	121.01	121.23	121.44
39	121.66	121.87	122.08	122.29
40	122.50	122.70	122.91	123.11

Table B–2
Male Height Factor

Inches	0.00	.25 (¼)	.50 (½)	.75 (¾)
60	75.23	75.35	75.48	75.60
61	75.72	75.84	75.96	76.09
62	76.21	76.33	76.45	76.56
63	76.68	76.80	76.92	77.04
64	77.15	77.27	77.39	77.50
65	77.62	77.73	77.84	77.96
66	78.07	78.18	78.30	78.41
67	78.52	78.63	78.74	78.85
68	78.96	79.07	79.18	79.29
69	79.40	79.50	79.61	79.72
70	79.83	79.93	80.04	80.14
71	80.25	80.35	80.46	80.56
72	80.67	80.77	80.87	80.98
73	81.08	81.18	81.28	81.38
74	81.48	81.58	81.68	81.78
75	81.88	81.98	82.08	82.18
76	82.28	82.38	82.47	82.57
77	82.67	82.77	82.86	82.96
78	83.05	83.15	83.24	83.34
79	83.43	83.53	83.62	83.72
80	83.81	83.90	83.99	84.09
81	84.18	84.27	84.36	84.45
82	84.54	84.64	84.73	84.82
83	84.91	85.00	85.08	85.17
84	85.26	85.35	85.44	85.53

B–8. Steps for preparing the Female Body Fat Content Worksheet, DA Form 5501–R, Dec 85

Name Print the soldier's last name, and middle initial in the NAME block. Also include her Rank, and Social Security Number.

Age Print her age in years in the AGE block.

Height Measure the soldier's height as described in this appendix, to the nearest quarter of an inch, and record the measurement in the HEIGHT block.

Weight Measure the soldier's weight as described in this appendix, to the nearest pound, and record in the WEIGHT block.

Note: Follow the rounding rules for rounding height and weight measurements as described earlier in this appendix.

Step 1. Weight Factor

Go to Table B–3, the Weight Factor Table, and locate the soldier's weight in the left-most column, which is in 10 pound increments. If the weight is exactly 120 pounds, the factor is found under the "0" column and is 147.24. If the weight is 121 pounds, the factor is found under the "1" column and is 147.62. If the weight is 126 the factor is found under the "6" column and is 149.47. Enter the appropriate weight factor in the CALCULATIONS section, STEP 11 A.

Step 2. Height Factor

Go to Table B–4, the Height Factor Table, and locate the soldier's height in the left-most column. If the height is a whole number, i.e., 64 inches, the factor is found under the 0.00 column and is 83.75. If the height is not a whole number, i.e., 64.25 inches, the factor is 84.07. If the height is 64.50 inches, the factor is 84.40, and if the height is 64.75 inches, the factor is 84.73. Enter the appropriate height factor in the CALCULATIONS section, STEP 11 D.

Step 3. Hip Measurement

Measure the soldier's hip circumference to the nearest quarter of an inch, and record in the block labeled "FIRST."

Step 4. Forearm Measurement

Measure the soldier's forearm to the nearest quarter of an inch, and record in the block labeled "FIRST."

Step 5. Neck Measurement

Measure the soldier's neck circumference to the nearest quarter of an inch, and record in the block labeled "FIRST."

Step 6. Wrist Measurement

Measure the soldier's wrist to the nearest quarter of an inch, and record in the block labeled "FIRST."

Note: REPEAT STEPS 3, 4, 5, and 6, IN SERIES, until you have completed 3 sets of Hip, Forearm, Neck and Wrist circumferences. When you have completed this series, find the mathematical average for *each* of the 4 circumference measures and place each average in its respective AVERAGE block.

Step 7. Hip Factor

Go to Table B–5, the Hip Factor Table, and locate the soldier's AVERAGE hip circumference in the left-most column. If the circumference is a whole number, i.e., 36 inches, the Hip Factor is found in the 0.00 column and is 15.83. If the circumference is not a whole number but is 36.25 inches, the factor is 15.94. If the circumference is 36.50 the factor is 16.05. Enter the appropriate factor in the CALCULATIONS section, 11 B.

Step 8. Forearm Factor

Go to Table B–6, the Forearm Factor Table, and locate the soldier's AVERAGE forearm circumference in the left-most column. If the circumference is a whole number, i.e., 10 inches, the factor is found under 0.00 column and is 39.97. If the circumference is not a whole number but is 10.25 inches, the factor is 40.97. If the circumference is 10.75 inches, the factor is 40.97. Enter the appropriate factor in the CALCULATIONS, 11 E.

Step 9. Neck Factor

Go to Table B–7, the Neck Factor Table, and locate the soldier's AVERAGE neck circumference in the left-most column. If the circumference is a whole number, i.e., 12 inches, the factor is found under the 0.00 column and is 16.25. If the circumference is not a whole number but is 12.25 inches, the factor is 16.59. If the circumference is 12.50 inches, the factor is 16.93. If the circumference is 12.75 inches, the factor is 17.26. Enter the appropriate factor is the CALCULATIONS section, 11 F.

Step 10. Wrist Factor

Go to Table B–8, the Wrist Factor Table, and locate the soldier's AVERAGE wrist circumference in the left-most column. If the circumference is a whole number, i.e., 7 inches, the factor is found under the 0.00 column and is 3.56. If the circumference is not a whole number but is 7.25 inches, the factor is 3.69. If the circumference is 7.50 inches, the factor is 3.82. If the circumference is 7.75 inches, the factor is 3.94. Enter the appropriate factor in the CALCULATIONS section, 11 G.

Calculations

Line C. Addition of Weight and Hip Factors

Add 11 A, Weight Factor, to 11 B, Hip Factor. Enter the result on line 11 C (Total).

Line H. Addition of Height, Forearm, Neck, and Wrist Factors

Add 11 D, Height Factor, 11 E, Forearm Factor, 11 F, Neck Factor, and 11 G, Wrist Factor together. Enter the result on line 11 H. (Total).

Line I. Percent Body Fat

Subtract Line 11–H from Line–C and enter on Line I. This is the soldier's PERCENT BODY FAT.

Table B–3
Female Weight Factor

Pounds	0	1	2	3	4	5	6	7	8	9
90	134.08	134.58	135.08	135.58	136.07	136.55	137.03	137.50	137.97	138.44
100	138.90	139.35	139.80	140.25	140.69	141.13	141.56	141.99	142.42	142.84
110	143.26	143.67	144.08	144.49	144.89	145.29	145.69	146.08	146.47	146.85
120	147.24	147.62	147.99	148.37	148.74	149.10	149.47	149.83	150.19	150.54
130	150.90	151.25	151.60	151.94	152.28	152.62	152.96	153.30	153.63	153.96
140	154.29	154.61	154.94	155.26	155.58	155.89	156.21	156.52	156.83	157.14
150	157.44	157.75	158.05	158.35	158.65	158.94	159.24	159.53	159.82	160.11
160	160.40	160.68	160.96	161.25	161.53	161.80	162.08	162.35	162.63	162.90
170	163.17	163.44	163.70	163.97	164.23	164.49	164.76	165.01	165.27	165.53
180	165.78	166.04	166.29	166.54	166.79	167.04	167.28	167.53	167.77	168.02
190	168.26	168.50	168.74	168.97	169.21	169.44	169.68	169.91	170.14	170.37
200	170.60	170.83	171.06	171.28	171.51	171.73	171.96	172.18	172.40	172.62
210	172.83	173.05	173.27	173.48	173.70	173.91	174.12	174.33	174.54	174.75
220	174.99	175.20	175.41	175.61	175.82	176.02	176.22	176.42	176.62	176.82
230	177.03	177.22	177.42	177.62	177.81	178.00	178.20	178.40	178.59	178.82
240	178.97	179.16	179.35	179.54	179.73	179.92	180.10	180.29	180.47	180.66
250	180.84	181.02	181.20	181.39	181.57	181.75	181.92	182.10	182.28	182.46
260	182.63	182.81	182.98	183.16	183.33	183.51	183.68	183.85	184.02	184.19
270	184.36	184.53	184.70	184.87	185.03	185.20	185.37	185.53	185.70	185.86

Table B–4
Female Height Factor

Inches	0.00	.25 (¼)	.50 (½)	.75 (¾)
55	71.97	72.30	72.62	72.95
56	73.28	73.61	73.93	74.26
57	74.59	74.91	75.24	75.57
58	75.90	76.22	76.55	76.88
59	77.20	77.53	77.86	78.19
60	78.51	78.84	79.17	79.49
61	79.82	80.15	80.48	80.80
62	81.13	81.46	81.78	82.11
63	82.44	82.77	83.09	83.42
64	83.75	84.07	84.40	84.73
65	85.06	85.38	85.71	86.04

Table B–4
Female Height Factor—Continued

Inches	0.00	.25 (¼)	.50 (½)	.75 (¾)
66	86.36	86.69	87.02	87.35
67	87.67	88.00	88.33	88.65
68	88.98	89.31	89.64	89.96
69	90.29	90.62	90.94	91.27
70	91.60	91.93	92.25	92.58
71	92.91	93.23	93.56	93.89
72	94.22	94.54	94.87	95.20
73	95.52	95.85	96.18	96.51
74	96.83	97.16	97.49	97.81
75	98.14	98.47	98.80	99.12
76	99.45	99.78	100.10	100.43
77	100.76	101.09	101.41	101.74
78	102.07	102.39	102.72	103.05
79	103.38	103.70	104.03	104.36
80	104.68	105.01	105.34	105.67
81	105.99	106.32	106.65	106.97
82	107.30	107.63	107.96	108.28
83	108.61	108.94	109.26	109.59
84	109.92	110.24	110.57	110.90

Table B–5
Female Hip Factor

Inches	0.00	.25 (¼)	.50 (½)	.75 (¾)
30	13.19	13.30	13.41	13.52
31	13.63	13.74	13.85	13.96
32	14.07	14.18	14.29	14.40
33	14.51	14.62	14.73	14.84
34	14.95	15.06	15.17	15.28
35	15.39	15.50	15.61	15.72
36	15.83	15.94	16.05	16.15
37	16.26	16.37	16.48	16.59
38	16.70	16.81	16.92	17.03
39	17.14	17.25	17.36	17.47
40	17.58	17.69	17.80	17.91
41	18.02	18.13	18.24	18.35
42	18.46	18.57	18.68	18.79
43	18.90	19.01	19.12	19.23
44	19.34	19.45	19.56	19.67
45	19.78	19.89	20.00	20.11
46	20.22	20.33	20.44	20.55
47	20.66	20.77	20.88	20.99
48	21.10	21.21	21.32	21.43
49	21.54	21.65	21.76	21.87
50	21.98	22.09	22.20	22.31
51	22.42	22.53	22.64	22.75
52	22.86	22.97	23.08	23.19
53	23.30	23.41	23.52	23.63

Table B–6
Female Forearm Factor

Inches	0.00	.25 (¼)	.50 (½)	.75 (¾)
5	19.98	20.98	21.98	22.98
6	23.98	24.98	25.98	26.98
7	27.98	28.98	29.98	30.97
8	31.97	32.97	33.97	34.97
9	35.97	36.97	37.97	38.97
10	39.97	40.97	41.97	42.97
11	43.96	44.96	45.96	46.96
12	47.96	48.96	49.96	50.96
13	51.96	52.96	53.96	54.96
14	55.95	56.95	57.95	58.95
15	59.95	60.95	61.95	62.95

Table B–7
Female Neck Factor

Inches	0.00	.25 (¼)	.50 (½)	.75 (¾)
5	6.77	7.11	7.45	7.79
6	8.12	8.46	8.80	9.14
7	9.48	9.82	10.16	10.49
8	10.83	11.17	11.51	11.85
9	12.19	12.53	12.86	13.20
10	13.54	13.88	14.22	14.56
11	14.90	15.23	15.57	15.91
12	16.25	16.59	16.93	17.26
13	17.60	17.94	18.28	18.62
14	18.96	19.30	19.63	19.97
15	20.31	20.65	20.99	21.33

Table B–8
Female Wrist Factor

Inches	0.00	.25 (¼)	.50 (½)	.75 (¾)
5	2.54	2.67	2.80	2.93
6	3.05	3.18	3.31	3.43
7	3.56	3.69	3.82	3.94
8	4.07	4.20	4.33	4.45
9	4.58	4.71	4.83	4.96
10	5.09	5.22	5.34	5.47
11	5.60	5.72	5.85	5.98
12	6.11	6.23	6.36	6.49
13	6.62	6.74	6.87	7.00
14	7.12	7.25	7.38	7.51
15	7.63	7.76	7.89	8.01

M TAB TAB TAB TAB TAB

BODY FAT CONTENT WORKSHEET *(Male)*

For use of this form, see AR 600–9; the proponent agency is DCSPER

NAME *(Last, First, Middle Initial)*	SSN	RANK	NOTE
DOE JOHN	198-00-0000	SGT	¼" = .25
HEIGHT *(to nearest 0.25 inch)*	WEIGHT *(to nearest pound)*	AGE	½" = .50
72	175	21	¾" = .75

STEP	FIRST	SECOND	THIRD	AVERAGE *(to nearest 0.25 in.)*
1. Measure abdomen at the level of the navel *(belly button)* to the nearest 0.25 inch. *(Repeat 3 times.)*	36.25	36.50	36.25	36.25
2. Measure neck just below level of larynx *(Adam's apple)* to the nearest 0.25 inch. *(Repeat 3 times.)*	16.25	16.25	16.50	16.25
3. Enter the average abdominal measurement to the nearest 0.25 inch.				36.25
4. Enter the average neck measurement to the nearest 0.25 inch.				16.25
5. Subtract Step 4 from Step 3 *(Enter result)* to the nearest 0.25 inch.				20.00
6. Find result from Line 5 *(the difference between Neck and Abdomen)* in Table B–1 *(Abdomen-Neck Factor)*. Enter factor.				99.48
7. Find the height in Table B–2 *(Height Factor)*. Enter factor.				80.67
8. Subtract Step 7 from Step 6 *(Enter result)*. This is Soldier's Percent Body Fat.				18.81

REMARKS

CHECK ONE

✓ Individual is in complaince with Army Standards; ____ is not in compliance with the standards.

____ Recommended monthly weight loss is 3–8 lbs.

PREPARED BY *(Signature)*	RANK	DATE	APPROVED BY SUPERVISOR *(Printed Name and Signature)*	RANK	DATE

DA FORM 5500–R, DEC 85

Figure B–1. Sample of a completed DA Form 5500

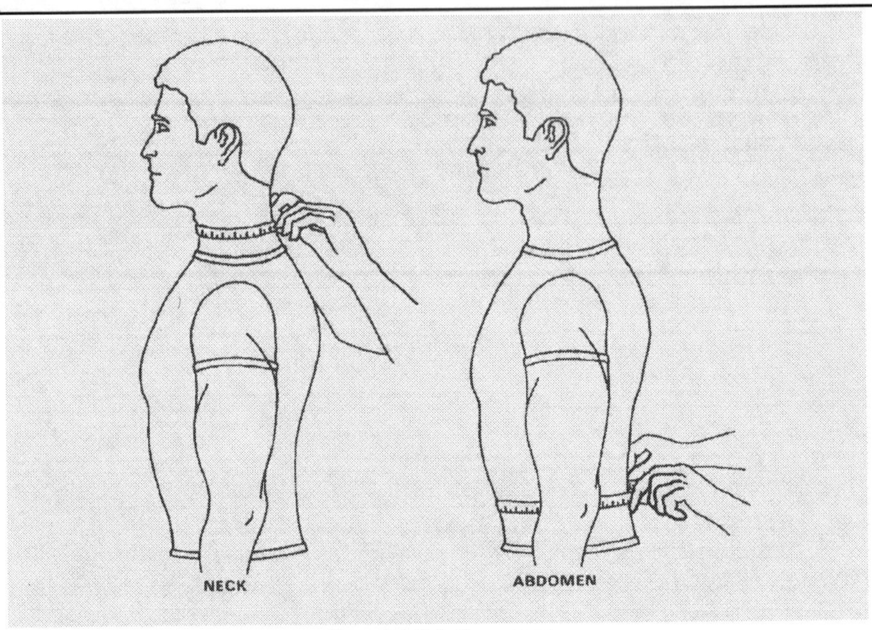

Figure B–2. Male Measurement

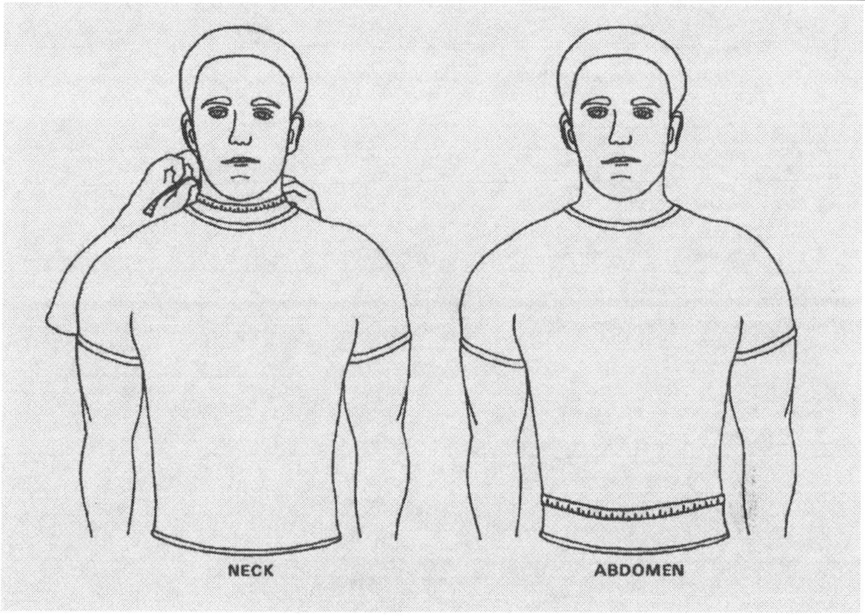

Figure B–2. Male Measurement—Continued

M — TAB — TAB — TAB — TAB

BODY FAT CONTENT WORKSHEET *(Female)*

For use of this form, see AR 600-9; the proponent agency is DCSPER

NAME *(Last, First, Middle Initial)*	SSN	RANK	NOTE:
DOE, JANE	432-00-0000	SGT	¼" = .25
HEIGHT *(to nearest 0.25 inch)*	WEIGHT *(to nearest pound)*	AGE	½" = .50
63.00	132	21	¾" = .75

STEP	FIRST	SECOND	THIRD	AVERAGE *(to nearest 0.25 in.)*
1. Find the soldier's weight in Table B-3 *(Weight Factor)*. Enter factor in 11A below				
2. Find soldier's height in Table B-4 *(Height Factor)*. Enter factor in 11D below.				
3. Measure hips at point where the gluteus muscles *(buttocks)* protrude backward the most. Round off to nearest 0.25 inch. Repeat three times, then average.	29.75	30.00	30.25	30.00
4. Measure forearm at its largest point *(with arm horizontal, palm up)* to nearest 0.25 inch. Repeat three times, then average.	8.25	8.50	8.75	8.50
5. Measure neck just below level of larynx *(Adam's apple)* to nearest 0.25 inch. Repeat three times and average.	13.25	13.50	13.75	13.50
6. Measure wrist between the bones of the hand and forearm to nearest 0.25 inch. Repeat three times, then average.	5.75	6.00	6.25	6.00
7. Find average hip measurement in Table B-5 *(Hip Factor)*. Enter factor in 11B below.				
8. Find average forearm measurement in Table B-6 *(Forearm Factor)*. Enter factor in 11E below.				
9. Find average neck measurement in Table B-7 *(Neck Factor)*. Enter factor in 11F below.				
10. Find average wrist measurement in Table B-8 *(Wrist Factor)*. Enter factor in 11G below.				

11. CALCULATIONS			REMARKS
A. Weight factor	151.60		
B. Hip factor	13.19		
C. TOTAL *(11A+11B)*		164.79	
D. Height factor	82.44		
E. Forearm factor	33.97		
F. Neck factor	18.28		
G. Wrist factor	3.05		
H. TOTAL *(11D+E+F+G)*		137.74	
I. SOLDIER'S PERCENT BODY FAT *(Line 11C—11H)*		27.05	

CHECK ONE

✓ Individual is in compliance with Army Standards; ____ is not in compliance with the standards.

____ Recommended monthly weight loss is 3–8 lbs.

PREPARED BY *(Signature)*	RANK	DATE	APPROVED BY SUPERVISOR *(Printed Name and Signature)*	RANK	DATE

DA FORM 5501-R, DEC 85

Figure B–3. Sample of a completed DA Form 5501

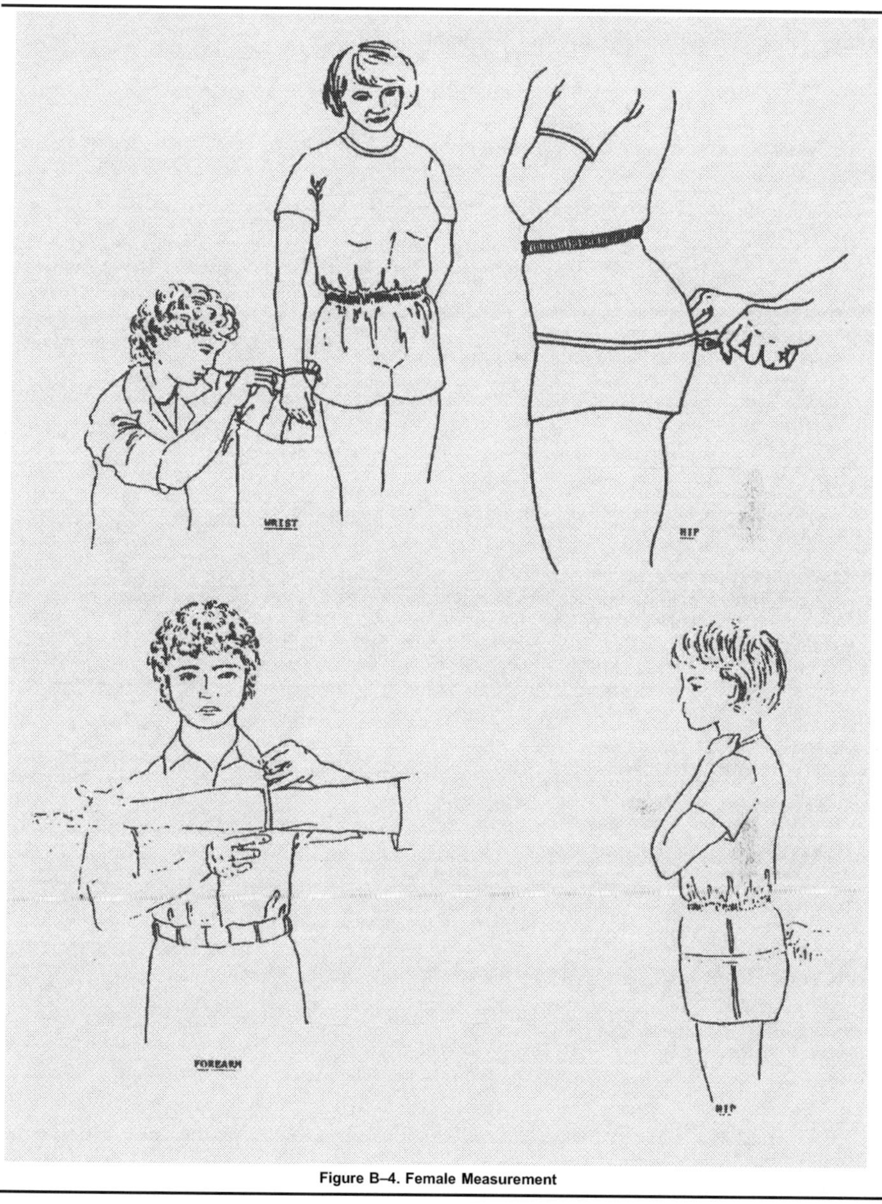

Figure B–4. Female Measurement

Appendix C
Nutrition Guide to the Weight Control Program

C–1. General
This appendix explains the basic principles of weight loss while maintaining normal nutrition. It does not replace the requirement for Active and Reserve Component Units with soldiers exceeding the body fat standards to be provided weight reduction counseling by qualified health care personnel. This guide should be used as a supplement to weight reduction counseling and as a guide for commanders in developing an effective weight control program.

C–2. Weight control—it's time to make the fitness connection
In many cases fitness begins with weight reduction. Aerobic fitness is related to an individual's body fat. The higher the fat, the less likely the individual is to be aerobically fit and the harder it is to maintain higher levels of physical stamina and endurance. A fit Army is a lean Army. Military readiness demands weight control—the spin-off is less likelihood of developing heart disease, high blood pressure, and diabetes. Proper nutrition and regular exercise are necessary to help you lose weight and improve your state of fitness.

a. Invest in yourself.

(1) Make a decision to lose weight and shape up.

(2) Get motivated.

(3) Develop a strategy (diet, exercise routine, lifestyle changes, etc.).

(4) Carry out the strategy.

b. Enjoy the payoff.

(1) A healthy appearance.

(2) An improved self image.

(3) A sense of accomplishment.

(4) A feeling of pride.

C–3. Making nutrition work
Improving your nutrition will increase your mileage in many ways. You can even lose weight while improving your performance. Your nutrition program should include the right number of calories to cause a steady loss of body fat with no loss of energy. Stay away from food fads—they're usually boring, unhealthy, too strict and lead only to temporary weight loss. Compare the benefits of a sound nutrition with the consequences of crash diets as shown below:

Table C–0–1
Nutrition

Sound Nutrition: Provides all required nutrients	**Crash Diets:** Most often lack some nutrients
Sound Nutrition: Gradual loss of body fat (1 to 2 pounds per week)	**Crash Diets:** Rapid loss of body water and muscle mass—not body fat
Sound Nutrition: Reinforces a good mental outlook	**Crash Diets:** Symptoms of grumpiness, headaches, anxiety and fatigue
Sound Nutrition: Improves health	**Crash Diets:** Can cause depression, dehydration, potential serious illness and a slow-down in your body's metabolism (the rate your body burns calories)
Sound Nutrition: Gives a sense of accomplishment	**Crash Diets:** End in eventual weight loss and failure
Sound Nutrition: Develops permanent good eating habits	**Crash Diets:** Encourage unhealthy eating habits of temporary duration

C–4. Basic strategy checklist—a plan for making the right connection

a. Good attitude. Having a good mental attitude is necessary to succeed in any program. To lose weight, a good mental attitude helps self-discipline—an important ingredient.

b. Sensible nutrition. A diet of adequate essential nutrients is necessary to prevent mental and physical fatigue. Crash diets don't work in a permanent weight control program.

c. Regular exercise. Exercise promotes physical fitness. It improves flexibility, strength, endurance and weight loss by speeding up the body's metabolism. It has also been proven to help supress the appetite.

d. Diet and exercise master plan. Create your own daily food intake and exercise plan. Keep a record. Make adjustments. You are in control.

e. Rest. Adequate rest improves attitude, posture, and appearance. Lack of rest and sleep can weaken resistance and will power.

f. A set goal. See a long range objective, then get going by setting easier-to-reach short-term goals. After the first few goals are met, the objective will be in plain view.

C–5. Balance check

If you need to lose weight, it's time to check your energy balance to see what's tilting the scale. When your energy input (calories) is greater than your energy output (activity), you store the extra calories as body fat. To stay in balance:

a. Increase exercise (frequency and intensity), or

b. Decrease calorie intake, or

c. Combine exercise with calorie reduction for best results.

C–6. Obesity risks

Excess body fat is harmful to your health. It increases your risk for developing high blood pressure, diabetes, heart disease, respiratory infections, gall bladder disease, low-back pain, and some forms of cancer. It has further drawbacks in physical appearance and interferes with physical performance.

C–7. Tailoring a nutrition program

The best nutrition program is one that allows you to lose body fat while you eat regular wholesome foods in controlled portion sizes. Does your diet include essential nutrients? Here's an easy way to check: (See below).

Table C–0–2
Diet structures

Food group	Recommended Servings Per day	Your diet
Milk (8 ozs)	2	—
Meat (3 ozs)	2	—
Bread or Cereal (1 slice or ½ cup)	4	—
Vegetables ½ cup)	2	—
Fruits (1 medium or 1/2 cup)	2	—
Does your diet provide the right amount of calories?		

Table C–0–3
Weight maintenance formula

________	X	________	=	________
Your ideal body weight		Activity factor		Your calorie level for maintenance

a. Weight maintenance.

b. Activity factors.

(1) *Sedentary.* Twelve to 14 calories per pound are required if you are not involved in exercise on the job or off duty.

(2) *Active.* Fifteen calories per pound are required when your job involves physical work and/or you are engaged in a regular exercise program.

(3) *Highly active.* Sixteen to 18 calories are required per pound when very physically demanding work and/or high level of physical training is done routinely. (Most people do not fit in this category).

c. Weight loss.

(1) To lose 1 pound of fat per week, subtract 500 calories per day from your calorie maintenance level.

(2) To lose 2 pounds of fat per week, subtract 1,000 calories per day. For example:

Table C–0–4
Weight loss formula

________________	less	________________	=	________________
Calorie level for maintenance. (A above)		Calories to meet your weekly goal. (B above)		Your daily calorie level for weight loss.

C–8. What about calories

Calories don't deserve a bad name unless your intake is greater than your requirement. Calories are provided by certain nutrients in the foods and beverages you consume.

a. The following nutrients provide calories:

Table C–0–5
Nutrient-calories guide

Nutrient: Carbohydrate (Provides an efficient fuel source for the body.)
Calories: 4 per gram

Nutrient: Protein (Provides material to repair and build tissues.)
Calories: 4 per gram

Nutrient: Fat (Provides essential fatty acids and concentrated energy source for the body.)
Calories: 9 per gram

Notes:
Alcohol, while not a nutrient does provide calories (7 per gram)

b. Listed in the table below are the nutrients that do not provide calories.

Table C–0–6
Vitamins–Minerals–Water

Nutrient: Vitamins (Needed to utilize the food you eat.)
Calories: 0

Nutrient: Minerals (Needed for bones, teeth, and chemical functions.)
Calories: 0

Nutrient: Water (Necessary for life.)
Calories: 0

C–9. Good nutrition—a personal choice

The amount of calories you consume depends on the type of food you choose, its preparation and the amount you eat (PORTION SIZE). Some foods are very concentrated in calories, (i.e., the portion size is small for the amount of calories it contains). Examples are fats, candy, fried foods, most deserts, and alcohol. These foods are also of low nutrient density—they provide few nutrients for the amount of calories provided. (See below for comparison.)

Table C–0–7
High calorie-low nutrient foods

	Calories
1 piece of pecan pie	750
1 large apple	80
One-half *fried* chicken breast	400
One-half *baked* chicken breast	180
1 potato, French *fried*	275
1 *baked* potato	100
ONE-HALF CUP *syrup pack* peaches	100
One-half cup *unsweetened* peaches	50

C–10. Easy ways to save calories

One sure way to save calories without reducing PORTION SIZE is to choose foods that are:

a. Lower is sugar. (Each teaspoon supplies an additional 20 calories).

b. Lower in fat. (Each teaspoon supplies an additional 45 calories).

c. See table C–1 and table C–2 for easy ways to save calories. The Right Image of Me in the Military (TRIMM) sample meals are lower in fat and sugar—thus more "nutrient dense" and lower in calories. See table C–3 for the 1200 calorie Daily Menu and table C–4 for the 1500-calorie Daily Menu.

C–11. Portion control

a. You'll need to learn to correctly estimate portion sizes in order to ensure adequate nutrition and to control intake. A guide to estimate portion sizes is shown below.
2 ounces=1 slice meat 3 x 4 inches and 1/4-inch thick (cooked)
3 ounces=1 meat pattie or portion (1/5 of a pound)
3-inch diameter, 1/2-inch thick (cooked)
3 ounces=1/2 small chicken breast
1 ounce=1 small chicken drumstick
2 ounces=1 chicken thigh
1 ounce=1/4 cup chopped meat (tuna, spagetti meatsauce, chili, ground meat)

b. Some equal measurements are as follows:

Table C–0–8
Equal measurements

1 cup	=	8 ounces
1 tablespoon	=	3 teaspoons
1/4 cup	=	4 tablespoons
1 ounce	=	40 grams
1 liter	=	34 ounces
.7 liter	=	24 ounces
.5 liter	=	17 ounces
.2 liter	=	7 ounces

c. There is no single food that is so high in calories that a small amount cannot be eaten occasionally. Many people, however, have a particular food obsession that must be recognized. For them trying to eat "just a cookie, piece of candy, or sparerib" is too tempting. The urge to eat "the whole thing" becomes too great. You have to make and follow your own rules according to your ability to control what you eat. Avoidance is one means of control. But if you plan your diet, and diet according to your plan, you can include a favorite high calorie food item as a special occasional treat.

C–12. "Good cookin' for good lookin'"— A memo to the cook for cutting calories during food preparation

a. The milk and cheese group.

(1) Use skim or lowfat milk in recipes when making puddings, sauces, soups, and baked products.

(2) Substitute plain, unsweetened lowfat yogurt or blenderized lowfat cottage cheese in recipes that call for sour cream or mayonaise.

b. The meat, poultry, fish, and dry beans group.

(1) TRIM fat from meat. Cook meats on rack so that fat can drain off.

(2) Roast, bake, broil, or simmer meat, poultry or fish without adding fat. Braise in covered pan on stove top or pan broil in a nonstick pan; and add spices to enhance flavors.

(3) Remove skin from chicken or turkey.

(4) Chill meat broth until fat turns light and solid on top. With a spoon or knife, skim or peel fat off and discard.

c. The vegetable and fruit group.

(1) Steam, boil, broil, or bake vegetables. Some fruits may be broiled or heated with spices added for flavor.

(2) Go easy on sauces, butter, and margarine. Season with herbs and spices. Crisp-cooked vegetables usually don't require as much seasoning as over-cooked vegetables.

(3) Try lemon juice or vinegar on salads. Cut way back on regular salad dressings. (One-fourth cup creamy dressing is approximately 340 calories!)

(4) Read nutrition information labels on food packages.

d. The bread and cereal group.

(1) Use less fat and sugar than called for in recipes. Substitute lower calorie ingredients.

(2) Avoid recipes for baked products that require large amounts of fat and sugar.

(3) Check ingredient labels for fat and sugar content. Check nutrition information label for total calories in each portion.

(4) Use diet margarine or plain yogurt on baked potatoes instead of margarine, butter, or sour cream.

(5) Have boiled, steamed, or baked rather than fried potatoes.

C–13. Dining tips

a. Avoid gravies, sauces, and deep-fried food. If the meat has been fried (southern style chicken, schnitzel), remove the coating and eat only the meat.

b. Remove all the visible fat from the meat.

c. Request diet salad dressing, vinegar, or lemon juice for your salad; most restaurants have them.

d. Starchy foods are not fattening when consumed in moderate quantities. However, avoid those prepared in cream sauces or deep-fried. For example, baked potatoes with a small amount of sour cream or margarine is a good choice. Also, noodles, rice, macaroni, or spaghetti are good potato substitutes.

e. Avoid rich desserts, ice cream, jello, pastry, candy, cookies, pies, cakes, sugar, honey, jam, jelly, regular soda, and other sweets. These are sources of concentrated calories that quickly cause your total intake to skyrocket in just a few bites. USE SPARINGLY, if you must.

f. If you MUST have a snack, have fresh fruits, a few crackers or pretzels, or delicious low-calorie raw vegetables.

g. Low-calorie beverages, black coffee, unsweetened sodas, or mineral water add no calories to your diet. Lowfat milk is also a nutritious choice. Carry individually packaged sugar substitutes to sweeten beverages. Try a slice of lemon or lime in a glass of ice water.

h. Alcohol does have calories as shown below:

Table C–0–9
Alcohol calories

	Calories
American beer, 12 oz.	160
European beer, 1/2 liter	250
Cocktails, 4–6 oz. (1/2 cup)	165
Hard liquor, 1½ oz. Jigger	110
Dry wine, .25 litre (8 oz.)	200
Sweet wine, .25 liter (8 oz.)	300

C–14. Educate your appetite

a. Follow these tips to lose weight and body fat.

(1) Eat S-L-O-W-L-Y

(2) Consume less fat.

(3) Take smaller portions

(4) Consume less sugar.

(5) Take smaller bites

(6) Chew food thoroughly

(7) Eat at least three regular meals per day.

(8) Plan snacks.

b. Plan your food intake and abide by your plan. Keep a food diary and monitor your own intake.

c. Always check your intake for balanced nutrition and total calories.

d. Become aware of how many calories you're consuming- especially in snack foods. This sample calorie count shows you why:

Table C–0–10
Sample calorie chart

	Calories
Chocolate milkshake (8 oz.)	840
Coke, soda, or sugared beverage, 12 ozs.	160
French fries, 20 pieces (2″ x 1/2″)	275
Fruit pie (1/6th of 9-inch pie)	410
Pecan pie (1/6th of 9-inch pie)	750
Cheesecake (1/6th of 9-inch pie)	800
Ice cream, 2 scoops	200
Potato or corn chips, 1 ounce package	180
Chocolate candy, 1 ounce	150
Grapefruit or orange juice, 1 cup	100
Big Mac	557
Quarter-pounder with cheese	521
AAFES Jumbo Cheeseburger	654
Cottage cheese, creamed, 1 cup	223
Beer (European), 1/2 liter	250
Creamy salad dressing, 1/4 cup	340
Salter nuts:	
peanuts, 2 tbs	170
cashews, 25 nuts	252
mixed, 25 nuts	188

C–15. Some final advice

a. Get motivated. Forget those old excuses: "But I've got a large frame—you know, big bones," or "Everyone in my family is big—it's hereditary,"or "Gosh, I haven't weighed 180 pounds since I was 15 years old,"or "I can pass the APFT, so why lose weight?"

b. If you're serious about losing weight, it will be easy to give up these typical excuses. Dieting is up to you—if you are willing to try. Don't be angry that you have to lose weight; think of it as something you have chosen to do. You may be a good soldier now, but you want to strive to be the best.

c. Try not to be food centered. Eating should be a source of enjoyment and satisfaction. It shouldn't be a substitute for feelings of boredom, anger, loneliness, or discouragement. Occupy your time with other activities not related to food. This is important if you plan on being successful at losing weight. Make up your mind to control food instead of letting food control you.

d. Construct new habit patterns. Make your meals last longer, slow down, enjoy flavors, pause halfway through meals and don't stuff yourself. Make meals as pleasant as possible even if you are cutting down on what you eat.

e. Don't be discouraged by weight plateaus (i.e., periods when no weight loss occurs despite your dieting and exercise efforts). Your body is adjusting to your new eating habits and changes are taking place. Stick with it!

f. You are always responsible for what you are eating. Don't cheat; remember, overeating hurts no one but yourself.

g. If you need extra help with your diet contact the nutrition clinic at your nearest treatment facility. For an exercise plan the physical therapy clinic at your nearest medical treatment facility is an excellent resource; master fitness trainers are also excellent resources to consult for basic nutrition information and exercise program development.

h. Use the attached "Personal Weight Loss Progress Chart" DA Form 5511–R which is located in the back of this handbook to keep track of your progress.

i. Remember all the work it took to get you to your desired weight. Don't let this be wasted effort.

j. For best results, combine dietary plans with regular exercise to:

k. Your dining facility provides low calorie menu plans to assist in your weight control efforts (see tables C–3 and C–4 for sample menu patterns).

BE☆ALL☆YOU☆CAN BE—NUTRITIONALLY!

Table C–1
Weight Control a Personal Choice

Instead of Amount	Calories	Substitute with	Calories
1 cup sweetened applesauce	232	unsweetened applesauce	100
3 oz. beef bologna	237	lean ham	103
1 oz. natural Swiss or cheddar cheese	110	part-skim milk mozzarella	90
1 cup cream-style cottage cheese	223	low-fat (1%) cottage cheese	125
1 oz. cream cheese	106	1 oz. Neufchatel cheese	70
1 cup vanilla ice cream	257	one-half cup vanilla ice cream	129
1 cup whole milk	170	skim milk	80
3 oz. dry salami	384	canned chicken	153
1 cup sour cream	454	low-fat yogurt	123
one-half cup frozen sweetened strawberries	139	frozen unsweetened strawberries	118
one-half cup oil-pack tuna	158	water-pack tuna	126
one-half cup syrup-pack canned fruit	90	drained/rinsed canned fruit	50
2 tablespoons mayonnaise	120	2 tablespoons mustard	24
4 tablespoons regular salad dressing	300	4 tablespoons diet salad dressing	80
1 piece fruit pie	410	1 piece fresh fruit	80

Notes:
REMEMBER: A calorie saved is a calorie burned...and 3,500 calories is equal to a pound of body fat.

Table C–2
Comparison of Regular and TRIMM* meals

Regular Food Item	Calories	TRIMM Food Item	Calories
Cream of tomato soup (8 oz.)	173	Tomato bouillon (8 oz.)	36
Meat loaf (4 oz.)	227	Meat loaf (3 oz.)	170
Brown Gravy (3 oz.)	130		
O'Brien potatoes	236	Parsley potatoes (1 med. 4½ oz.) NO BUTTER	99
Club spinach (w/eggs) ½ cup	153	Spinach (w/lemon) ½ cup	29
Tossed green salad (1 cup)	10	Tossed green salad (1 cup)	10
1000 Island Dressing (2 tbsp. or 1 oz.)	165	Diet 1000 Island Dressing (2 tbsp. or 1 oz.)	54
Yellow cake, chocolate frosting (3 inch squares.)	365	Fresh fruit (1 piece)	80
Milk, whole (8 fl. oz.)	159	Milk, skim (8 fl. oz.)	80
Totals	1618		558

Notes:
* TRIMM=The Right Image of Me in the Military.
The TRIMM Sample Meal is lower in fat and sugar—thus more "nutrient dense" and lower in calories

Table C–3
Sample 1200 Calorie Menu

	Average Calories
Breakfast	
1 cup fruit or juice (unsweetened)	100
1 ounce meat, cheese or egg (prepared without added fat)	80
1 slice toast (white, whole wheat, rye, or ¾ cup dry cereal)	70
1 cup skim milk	80
Total	330
Lunch	
2-ounce serving lean meat (or 1 oz. meat and 1 oz. cheese)	160
2 slices of bread or 1 bun or roll	140
1 serving cooked vegetables (leaves and stems—25 calories per ½ cup; starchy vegetables corn, lima beans—70 calories per one-third cup)	25
1 serving tossed vegetable salad	10
Vinegar or lemon juice	—
1 piece of *small* fresh fruit (or ¼ cup drained canned fruit)	50
Total	385
Supper	
3-ounce serving lean meat (no gravy)	225
one-half cup starch (potato, rice, pasta, dry beans or starchy vegetables)	80
1 serving cooked vegetables	25
1 serving tossed vegetable salad	10
Vinegar or lemon juice	—
1 piece of *small* fresh fruit (or ¼ cup drained canned fruit)	50
1 cup skim milk	80
Total	470
Choose Zero Calorie Beverages:	
Coffee or tea without cream or sugar, club soda, mineral water, iced water with lemon or lime.	0
Approximate Total for Day	1185

Table C–4
Sample 1500—Calorie Menu

	Average Calories
Breakfast	
1 cup fruit or juice (unsweetened)	100
1 ounce meat, cheese or egg, prepared without fat	80
1 slice toast (white, whole wheat, rye)	70
One-half cup cooked cereal or ¾ cup dry cereal (unsweetened)	70
1 cup skim milk	80
1 teaspoon margarine	45
Total	445
Lunch	
3 oz. serving lean meat (or 2 oz. meat and 1 slice cheese)	210
2 slices of bread or 1 bun or 1 roll	140
1 serving cooked vegetables (Leaves and stems 25 calories per ½ cup; starchy vegetables 70 calories per one-third cup)	25
1 tossed vegetable salad	10
1 tablespoon low-calorie dressing	30
1 piece of fresh fruit (medium to large)	80
1 small or ¼ cup drained fruit, 40 calories	
Total	520
Dinner	
3 oz. serving lean meat	225
one-half cup starchy vegetable (with small amount added fat)	100
Serving cooked vegetable	25
1 tossed vegetable salad	10
1 tablespoon low-calorie dressing	30
1 piece fresh fruit (medium to large)	80
1 cup skim milk	80
Total	550
Choose Zero Calorie Beverages:	
Coffee or tea without cream or sugar, club soda, mineral water, ice water with lemon or lime.	0
Approximate Total for Day	1515

Glossary

Section I
Abbreviations

AD
active duty

ADS
active duty support

ADT
active duty for training

AGR
Active Guard/Reserve

APFT
Army Physical Fitness Test

ARNG
Army National Guard

ARPERCEN
US Army Reserve Personnel Center

CAR
Chief, Army Reserve

CG
commanding general

CNGB
Chief, National Guard Bureau

DA
Department of the Army

DCSPER
Deputy Chief of Staff for Personnel

DCSLOG
Deputy Chief of Staff for Logistics

DOD
Department of Defense

FORSCOM
US Army Forces Command

HQDA
Headquarters, Department of the Army

HSC
US Army Health Services Command

IRR
Individual Ready Reserve

MACOM
major Army command

RC
Reserve Component

SADT
Special Active Duty for Training

TPU
troop program unit

TRADOC
US Army Training and Doctrine Command

TSG
The Surgeon General

USAR
US Army Reserve

Section II
Terms

Body composition
The human body is composed of two major elements: lean body mass (which includes muscle, bone, and essential organ tissue) and body fat. Body fat is expressed as a percentage of total body weight that is fat. For example, an individual who weighs 200 pounds and 18 percent body fat has 36 pounds of fat. Women generally have a higher percentage of body fat than men because of genetic and hormonal differences; thus, body fat standards differ among men and women by age groups.

Health care personnel
Trained physicians (military or civilian employees or contract personnel), physician's assistants, Registered Nurses, dieticians, and physical/occupational therapists under supervision of the unit surgeon or the commander of the medical treatment facility. For the purpose of this regulation, this term includes personnel of US forces and host nations.

Overweight
A soldier is considered overweight when his or her percent body fat exceeds the standard specified in paragraph 20*c* of this regulation.

Professional military or civilian schooling
This includes all individual training courses beyond Initial Entry Training (IET). It does not include unit training involving crews and teams. IET includes basic branch course or equivalent for officers; warrant officer entry course for non-prior service personnel; and basic training, AIT, OSUT, and OST for enlisted personnel.

Satisfactory progress
Progressing toward a point to meet the body fat standards described in paragraph 19*c*. Weight loss of 3–8 pounds per month is required for satisfactory progress.

Weight control
An individual program by which each soldier attains and maintains an acceptable weight and body composition through self-motivation or involvement in an official weight control program.

Section III
Special Abbreviations and Terms
There are no special terms.